REJECTING NATURALISTIC THEORIES OF ORIGINS

SCIENTIFIC AND SCRIPTURAL ARGUMENTS

Thomas N. Thrasher, Ed.D., Ph.D.

Thrasher Publications

Copyright © 2019 Thomas N. Thrasher

All rights reserved.

ISBN: **9781099739866**

DEDICATION

To All Who, Having Seen the Inadequacies

And Errors of Evolutionary Theory, Have Come to

Appreciate the Magnificence of the Creator (Genesis 1:1)

CONTENTS

	Introduction	i
1	Defining the Issue	3
2	Arguments Based on Scripture	25
3	Arguments Based on Science	43
4	Conclusions and Recommendations	84

INTRODUCTION

Evolutionary theory and speculation dominate much of our world today. One can scarcely watch movies or television programs, read magazines or newspapers, visit amusement parks or museums, or attend school without being bombarded, sometimes subtly, at other times explicitly and unquestioningly, with the idea of the evolution of all things by entirely naturalistic processes over a very long period of time.

Consequently, one of the most fundamental and important issues confronting people in today's world is deciding what is the most rational explanation for the origin and development of all things. Numerous arguments and sources of evidence have been submitted in favor of divergent viewpoints. This book presents both scriptural and scientific evidence for rejecting the wholly naturalistic view of origins in favor of the view that God created the universe by supernatural actions.

The bulk of the material presented in this book was originally the body of a dissertation submitted in partial fulfillment of the requirements for the Doctor of Philosophy degree in Christian Apologetics, awarded in 2015.

CHAPTER ONE

DEFINING THE ISSUE

One of the most fundamental and important issues debated in today's world is deciding what is the most rational explanation for the origin and development of all things. Numerous arguments and sources of evidence have been submitted in favor of divergent viewpoints. This dissertation presents both scriptural and scientific evidence for rejecting the wholly naturalistic, gradualistic view of origins and accepting that God created the world as described in the Bible.

In essence there are only two explanations for the existence of the universe and the host of things it contains: divine creation or naturalistic evolution. Of course, these two general categories may be subdivided into numerous subcategories. For example, there are differing viewpoints of creation, often depending upon the particular culture or civilization and its conception of God or gods. Some take a strict, biblical view that God made the world in six literal days (which, by the way, is the view supported in this document), while others allow for long periods of time, even millions or billions of years, in which God brought all things into existence.

Similarly, many variations of the naturalistic, evolutionary perspective may be found in the literature. For example, some have held that the universe in some form has always existed—it is eternal, while many today contend that the universe originated in an enormous explosion, usually called the Big Bang, perhaps 15 billion years ago, although the age of the universe is variously dated by those who hold an evolutionary perspective. Although there are many differing views on evolution and its details, in this dissertation the term "theory of evolution" will sometimes be used, since that is a frequently employed designation. However, it should be observed that there are actually numerous theories (plural) seeking to set forth a naturalistic alternative to divine creation.

Defining "Evolution"

The term "evolution" is used in different ways. Smallwood and Green (1971) explain:

> You must be aware of the dual usage of the term 'evolution.' Literally it means 'change.' When the term is applied to species, we have enough evidence to say it is a fact that evolution has occurred and does occur. There is also the theory of evolution, which states that all of our modern species are the modified descendants of species that lived in the past. Moreover, the theory states that all present-day species have evolved from ancestors that were originally formed under natural conditions. Therefore, it is important to differentiate between the 'fact' and the 'theory' of evolution. (p. 190)

Mader (2010) provides the following definition for "evolution": "Descent of organisms from common ancestors with the development of genetic and phenotypic changes over time that make them more suited to the environment" (p. G-10). Of course, when this dissertation uses the term evolution, mere change is not intended. Instead, evolution refers to what is sometimes called *macroevolution*, "large-scale evolutionary change, such as the formation of new species" (Mader, p. G-16). This includes a progression from nonliving things to human beings over billions of years entirely by means of naturalistic processes. The theory of evolution does not credit God for anything throughout the process.

Although many science writers fail to emphasize that the theory of evolution is a theory, not a fact, Smallwood and Green are careful to make this distinction: "When we use the term 'evolution' to describe a sequence of changes that originated life and transformed it into the populations that we know today, we must keep in mind that this is a theory—the theory of evolution" (p. 186).

Feldkamp (2002) explains what a "theory" is:

In science, a theory may be formed after many related hypotheses have been tested and supported with much experimental evidence. The word *theory* does not mean "wild guess"; it does not even mean "hypothesis." Rather, a **theory** is a broad and comprehensive statement of what is thought to be true.

A theory is supported by considerable evidence and may tie together several related hypotheses. Few true theories are produced in science, relative to the volume of work performed and the number of hypotheses tested. (p. 19)

Despite the claims of numerous evolutionists, naturalistic evolution is merely a theory, not a scientific fact. An old story tells of someone asking an audience the question: "How many legs does a normal dog have?" The audience responded, "Four." The questioner then asked, "If we call the tail a leg, how many legs would the dog have?" Someone said, "Five!" The questioner than explained, "No. There would still be four. Calling the tail a leg doesn't make it so!" Similarly, calling naturalistic evolution a fact does not make it so.

Emphasizing the vast extent to which evolution has supposedly occurred during billions of years of earth's history, Hotton (1968) explains:

> For the history of all life on earth could as well be seen as one continuous and arduous quest for sustaining energy, and it has been principally as a means of succeeding in that quest, that life, over billions of years, has 'evolved' from a microscopic bit of sea-borne jelly to more than 1.25 million different species ranging from asters to zebras, and including you and me. (p. 7)

This represents the commonly announced "theory," but it is not a scientific fact.

Alternative Explanations

Although claims have been made by some advocates of evolution that it is a scientific fact, there are actually alternative explanations. Lipson (1980) stated:

> If living matter is not, then, caused by the interplay of atoms, natural forces and radiation, how has it come into being? There is another theory, now quite out of favour, which is based upon the ideas of Lamarck: that if an organism needs an improvement it will develop it, and transmit it to its progeny. I think, however, that we must go further than this and admit that the only acceptable explanation is *creation*. I know that this is anathema to physicists, as indeed it is to me, but we must not reject a theory that we do not like if the experimental evidence supports it. (p. 138)

Consequently, creation is an alternative to naturalistic

evolution that should be considered on the basis of the available evidence, whether positively in favor of creation or negatively against evolution. Evolution is not the only option.

The Teaching of Evolution is Controversial

The teaching of evolution, particularly to high school students, has been met with strong opposition. For example, several years ago the Alabama State Board of Education affixed notices to the inside front cover of biology textbooks. An excerpt of this message reads:

> This textbook discusses evolution, a controversial theory some scientists present as a scientific explanation for the origin of living things, such as plants, animals and humans.
>
> No one was present when life first appeared on earth. Therefore, any statement about life's origins should be considered as theory, not fact.
>
> The word "evolution" may refer to many types of change…. Evolution may … refer to the change of one living thing to another, such as reptiles into birds…. Evolution also refers to the unproven belief that random, undirected forces produced a world of living things.
>
> Pro-evolution advocates called for these notices to be

removed.

Admitting the possibility that "scientific truth" may be in error, Otto, Towle, and Bradley (1981) state:

> In the Middle Ages, "scientific truths" were handed down from earlier times. These "truths" were not to be questioned, much less disapproved. People who dared to disagree were ridiculed, attacked, and sometimes killed. Even a generation ago, since was believed to be much more exact than it is now known to be. Today, people no longer think of any scientific answer as a final one. They know that ideas may have to change in the light of new discoveries. (p. 4)

Eiseley (1957) recognized that scientists had developed their own mythology with reference to evolutionary theory:

> With the failure of these many efforts science was left in the somewhat embarrassing position of having to postulate theories of living origins which it could not demonstrate. After having chided the theologian for his reliance on myth and miracle, science found itself in the unenviable position of having to create a mythology of its own: namely, the assumption that what, after long effort, could not be proved to take place today had, in truth, taken place in the primeval past. (p. 199)

Popularity of Evolution

The evolutionary approach was not originated by Charles Darwin, but through the publication of his celebrated book *On the*

Origin of Species by Means of Natural Selection, or the Preservation of Favoured Races in the Struggle for Life (1859), a title usually shortened to *The Origin of Species*, he certainly popularized the theory of evolution. In *The Biblical Basis for Modern Science* (Morris, 2002), John W. Oller states: "During Darwin's heyday, in the 19th century, it became popular to suppose that the material things and living beings in the real world could come about by pure chance and without any assistance whatever from God" (p. 9).

Feldkamp (2002) describes the beginnings of the modern theory of evolution:

> In the mid-1800s, both Charles Darwin (1809-1882) and Alfred Wallace (1823-1913) independently proposed the hypothesis that species were modified by natural selection.... While Darwin and Wallace announced their hypotheses at the same time, Darwin's name became more associated with evolutionary theory after his book *The Origin of Species* was published in 1859. (p. 283)

Describing Darwin's fundamental ideas, Feldkamp (2002) says: "Darwin's ideas about evolution and natural selection are summed up in two theories. In his first theory, Darwin stated that evolution, which is defined as descent with modification, occurs. In his second theory, Darwin proposed that a process he called natural selection causes evolution" (p. 286). Regarding Darwin's view of the origin of the various species that have existed, Feldkamp says, "Darwin inferred that *all* species had descended from one or a few original types of life" (p. 286).

Hotton (1968) writes of the prominent role of Darwin:

> Modern evolutionary theory was first set forth in detail by an English naturalist named Charles Robert Darwin…. Since then evolutionary thought has played a prominent, and frequently controversial, role in changing man's view of his universe from a static one, in which he is the apex of nature, to one that is dynamic, in which he occupies a small and undoubtedly transient place. (p. 7)

Although evolution, as used in this document, is only a theory espoused by many people today, it is often claimed that evolution is a fact. Dr. Archie L. Manis, who was at the time an Associate Professor of Biology at Abilene Christian University, states: "Evolution's history and methodology will continue to feed debates for generations, but the fact of evolution is beyond dispute. The concept is rational, scientific, and supported by an overwhelming mass of evidence from past and present" (quoted by Dr. Bert Thompson in *Is Genesis Myth?*, 1986, p. 77). Similarly, we read the claim that "evolution is a fact, not a theory. It once was a theory, but today, as a consequence of observation and testing it is probably the best authenticated actually known to science. There are theories concerning the mechanisms of evolution, but no competent student doubts the reality of evolution" (Montagu, 1984, p. 13). Dobzhansky (1955) expressed a similar idea: "Man is ... a product of a long evolutionary development ... the evidence shows conclusively that man arose from forebears who were not men" (p. 319).

A primary reason that evolution is so popular is that many reject the existence of God. Some form of evolutionary philosophy is the only real alternative. D. M. S. Watson (1929) frankly stated, "Evolution [is] a theory universally accepted not because it can be proven by logically coherent evidence to be true, but because the only alternative, special creation, is clearly incredible" (p. 233)

Richard Lewontin (1997) admitted:

We take the side of science *in spite* of the patent absurdity of some of its constructs, *in spite* of its failure to fulfill many of its extravagant promises of health and life, *in spite* of the tolerance of the scientific community for unsubstantiated just-so stories, because we have a prior commitment, a commitment to materialism. It is not that the methods and institutions of science somehow compel us to accept a material explanation of the phenomenal world, but, on the contrary, that we are forced by our *a priori* adherence to material causes to create an apparatus of investigation and a set of concepts that produce material explanations, no matter how counter-intuitive, no matter how mystifying to the uninitiated. Moreover, that materialism is an absolute, for we cannot allow a Divine Foot in the door. (p. 31).

Organizations such as the National Association of Biology Teachers (NABT) have issued statements expressing the viewpoint that evolution is a fact to the degree that they support teaching no other approach. The *NABT Position Statement on Teaching Evolution* (2011) states:

The frequently-quoted declaration of Theodosius Dobzhansky (1973) that "Nothing in biology makes sense except in the light of evolution" accurately reflects the central, unifying role of evolution in the science of biology. As such, evolution provides the scientific framework that explains both the history of life and the continuing change in the populations of organisms in response to environmental challenges and other factors. Scientists who have carefully evaluated the evidence overwhelmingly support the conclusion that both the principle of evolution itself and its mechanisms best explain what has caused the variety of organisms alive now and in the past.

Sarfati (1999) correctly characterizes the state of secular education, particularly in the United States: "The whole secular education system in America (and most other countries around the world) is underpinned by evolution. After reviewing a number of biology textbooks in the secular school system, we find they are *all* blatantly pro-evolution" (p. 13).

With reference to the alleged biases of creationists as contrasted with the openness of evolutionists, Sarfati (1999) explains:

> Creationists often appeal to the facts of science to support their view, and evolutionists often appeal to philosophical assumptions from outside science. While creationists are often criticized for starting with a bias, evolutionists also start with a bias, as many of them admit. The debate between creation and evolution is primarily a dispute between two

world views, with mutually incompatible underlying assumptions." (p. 15)

Sarfati observes an often overlooked fact of history: "Most branches of modern science were founded by believers in *creation*" (p. 25). The following table is only a short summary illustrating the reality of his statement.

Branch of Science	Examples of Creationist Scientists
Astronomy	Copernicus, Galileo, Herschel, Kepler, Maunder
Biology	Agassiz, Linnaeus, Mendel, Pasteur, Ray, Virchow
Chemistry	Boyle, Dalton, Ramsay
Geology	Brewster, Buckland, Cuvier, Steno, Woodward
Mathematics	Euler, Leibnitz, Pascal
Physics	Faraday, Kelvin, Maxwell, Newton

Theistic Evolution

Many people attempt to occupy a middle-of-the-road stance commonly called theistic evolution. Henry Morris (2002) describes them by writing: "There are multitudes of professing Christian people who think they can believe both the Bible and evolution—that evolution is merely God's method of creation…. [However,] anyone who believes this … simply does not understand either evolution or the Bible or both" (p. 97). Morris continues by providing five reasons why evolutionary theory and the Bible cannot be harmonized.

1. Genesis 1 states 10 times that God's arrangement from the time of creation is that each living thing reproduces "according to its kind." Although God's arrangement permits some variation within the created kinds, it precludes macroevolution of one kind into another, as demanded by the naturalistic theory of evolution.

2. God's work of creation ended at the conclusion of the six days of the creation week: "Thus the heavens and the earth, and all the host of them, were finished. And on the seventh day God ended His work which He had done, and He rested on the seventh day from all His work which He had done…. He rested from all His work which God had created and made" (Genesis 2:1-3).

"Consequently, present-day biological processes (variation, mutation, even speciation) could not be processes of creation or development, as theistic evolutionists must allege" (Morris, p. 97).

3. At the conclusion of the six days of creation, God pronounced that everything He had made was "very good": "Then God saw everything that He had made, and indeed *it was* very good. So the evening and the morning were the sixth day" (Genesis 1:31). This description is "inconsistent with a system of nature ruled by tooth and claw, a grinding struggle for existence, with only the fittest and more prolific surviving" (p. 97).

4. The teaching of Jesus, "who is himself the Creator of all things (John 1:3), ... [was] that the Genesis record of creation ... was intended to be taken historically and literally (see Matt. 19:4-6; Mark 10:6-9)" (p. 97). For example, Jesus declared, "But from the beginning of the creation, God 'made them male and female.'" (Mark 10:6). God did not make man and woman after billions of years. He made them "from the beginning of the

creation," in fact, on the sixth day of the creation week. Jesus Himself endorsed that historical occurrence.

5. "Evolution is the most wasteful, inefficient, and heartless process that could ever be devised by which to produce man. If evolution is true, then billions upon billions of animals have suffered and died in a cruel struggle for existence for a billion years…. The God of the Bible could never be guilty of such a cruel and pointless charade as this!" (p. 97).

In addition to these points made by Morris, additional reasons involve conflicts between evolutionary theory and the Bible. For example, according to some evolutionists, "birds are thought to have evolved from small, fast-running carnivorous dinosaurs during the Jurassic period (208-144 million years ago)" (Feldkamp, p. 862). However, according to the account in Genesis 1, birds were created on the **fifth day** of creation: "Then God said, "… let birds fly above the earth across the face of the firmament of the heavens." So God created … every winged bird according to its kind. And God saw that *it was* good…. So the evening and the morning were the fifth day" (Genesis 1:20-23). Land beasts were not created until the **sixth day**: "And God made the beast of the earth according to its kind, cattle according to its kind, and everything that creeps on the earth according to its kind. And God saw that *it was* good.... Then God saw everything that He had made, and indeed *it was* very good. So the evening and the morning were the sixth day." (Genesis 1:25, 31). How could birds have evolved during millions of years from certain land animals if birds were created **before** such animals (Genesis 1)?

Each creation day had an evening and a morning (Genesis 1:5, 8, 13, 19, 23, 31); therefore, the days were not long periods of time. If the Genesis account of creation is true, and God created all things during a period of six days, then evolution cannot possibly have occurred, since it requires very long periods of time.

Man was made in the image of God, not gradually evolved from some lower forms of life (Genesis 1:26-27; 5:1): "Man was created in the image and likeness of God. The Bible clearly states that man was formed by the hand of the Creator Himself" (Riegle, 1971, p. 87). What is sometimes designated as the "molecules to man" theory (gradual evolution over a very long period of time) is contradictory to the Biblical record.

Clearly, the teaching of the Bible is contradictory to the teaching of this aspect of evolutionary theory. Both cannot be true.

The Bible teaches that man is different from animals in significant ways: "Then God said, 'Let Us make man in Our image, according to Our likeness; let them have dominion over the fish of the sea, over the birds of the air, and over the cattle, over all the earth and over every creeping thing that creeps on the earth.' So God created man in His *own* image; in the image of God He created him; male and female He created them. Then God blessed them, and God said to them, 'Be fruitful and multiply; fill the earth and subdue it; have dominion over the fish of the sea, over the birds of the air, and over every living thing that moves on the earth'" (Genesis 1:26-28). Man, unlike any of the animals, was made "in the image of God." Furthermore, when considering a mate for man, God decided that the various animals were unsuitable, so He made woman (Genesis 2:18-25).

The Bible teaches that man was created as man on the same creation day as land beasts (the sixth day of creation), not that man evolved from such animals over millions of years. Man was fully man on the day God made him. Man and woman possessed an ability to communicate using language; both could understand what God said (Genesis 2:16-17; 3:1-4); both could speak intelligibly (Genesis 3:2-3, 10ff); both were accountable to God for their actions (Genesis 3:16-24). One cannot fully believe the Bible and hold the theory of gradual, naturalistic evolution.

Organization of the Study

This dissertation presents both scriptural and scientific arguments for rejecting a naturalistic explanation for the origin and development of what we observe in the universe. Our attention will be focused on arguments that are inconsistent with or contradictory to naturalistic, materialistic evolutionary theories. In Chapter 2 scriptural arguments supporting the view that God created the universe and all things in it during a period of six days will be considered. In Chapter 3 the focus will be on numerous scientific arguments that undermine the view that all living things have resulted from a gradual, materialistic, naturalistic process that occurred during billions of years. Chapter 4 will offer conclusions and recommendations related to the evidence presented.

CHAPTER 2

ARGUMENTS BASED ON SCRIPTURE

God Created the Universe

The Scriptures repeatedly point out that the world was *created* and that God was the *Creator*. Although all relevant Bible references will not be cited in order to conserve space, a sufficient number will be provided to establish indisputably that the Bible teaches the creationist view. Biblical quotations are from the New King James Version unless otherwise specified. Bold print that highlights some terms in the original quotations is added for emphasis.

Old Testament Affirmation

Genesis 1:1—"In the beginning God **created** the heavens and the earth."

Genesis 1:21—"So God **created** great sea creatures and every living thing that moves, with which the waters abounded, according to their kind, and every winged bird according to its kind. And God saw that *it was* good."

Genesis 1:27—"So God **created** man in His *own* image; in the image of God He **created** him; male and female He **created** them."

Genesis 2:3-4—"Then God blessed the seventh day and sanctified it, because in it He rested from all His work which God had **created** and made. This *is* the history of the heavens and the earth when they were **created**, in the day that the Lord God made the earth and the heavens."

Genesis 5:1-2—"This is the book of the genealogy of

Adam. In the day that God **created** man, He made him in the likeness of God. He **created** them male and female, and blessed them and called them Mankind in the day they were **created**."

Genesis 6:7—"So the Lord said, "I will destroy man whom I have **created** from the face of the earth, both man and beast, creeping thing and birds of the air, for I am sorry that I have made them."

Deuteronomy 4:32—"For ask now concerning the days that are past, which were before you, since the day that God **created** man on the earth, and *ask* from one end of heaven to the other, whether *any* great *thing* like this has happened, or *anything* like it has been heard."

Psalm 104:30—"You send forth Your Spirit, they are **created**; And You renew the face of the earth."

Psalm 148:5—"Let them praise the name of the Lord, For He commanded and they were **created**."

Ecclesiastes 12:1—"Remember now your **Creator** in the

days of your youth, Before the difficult days come, And the years draw near when you say, 'I have no pleasure in them.'"

Ecclesiastes 12:6—"Remember your **Creator** before the silver cord is loosed, Or the golden bowl is broken, Or the pitcher shattered at the fountain, Or the wheel broken at the well."

Isaiah 40:26—"Lift up your eyes on high, And see who has **created** these *things,* Who brings out their host by number; He calls them all by name, By the greatness of His might And the strength of *His* power; Not one is missing."

Isaiah 40:28—"Have you not known? Have you not heard? The everlasting God, the Lord, The **Creator** of the ends of the earth, Neither faints nor is weary. His understanding is unsearchable."

Isaiah 41:20—"That they may see and know, And consider and understand together, That the hand of the Lord has done this, And the Holy One of Israel has **created** it."

Isaiah 42:5—"Thus says God the Lord, Who **created** the heavens and stretched them out, Who spread forth the earth and that which comes from it, Who gives breath to the people on it, And spirit to those who walk on it."

Isaiah 45:12—"I have made the earth, And **created** man on it. I—My hands—stretched out the heavens, And all their host I have commanded."

Isaiah 45:18—"For thus says the Lord, Who **created** the heavens, Who is God, Who formed the earth and made it, Who has established it, Who did not create it in vain, Who formed it to be inhabited: 'I *am* the Lord, and *there is* no other.'"

These Old Testament passages clearly establish that God is Creator. The Hebrew word *bārā'* (to create, make) is discussed in *Nelson's Expository Dictionary of the Old Testament*: "This verb is of profound theological significance, since it has only God as its subject. Only God can 'create' in the sense implied by *bārā'*. The verb expresses creation out of nothing, an idea seen clearly in passages having to do with creation on a cosmic scale. (p. 51; published as a component of *Vine's Expository Dictionary of Biblical Words*).

New Testament Affirmation

Among the New Testament passages affirming creation and God as Creator are the following:

Mark 10:6—"But from the beginning of the **creation**, God 'made them male and female.'"

Mark 13:19—"For *in* those days there will be tribulation, such as has not been since the beginning of the **creation** which God created until this time, nor ever shall be."

Romans 1:20—"For since the **creation** of the world His invisible *attributes* are clearly seen, being understood by the things that are made, *even* His eternal power and Godhead, so that they are without excuse."

Romans 1:25—"who exchanged the truth of God for the

lie, and worshiped and served the creature rather than the **Creator**, who is blessed forever. Amen."

Romans 8:19—"For the earnest expectation of the **creation** eagerly waits for the revealing of the sons of God."

Romans 8:20-22—"For the **creation** was subjected to futility, not willingly, but because of Him who subjected *it* in hope; because the **creation** itself also will be delivered from the bondage of corruption into the glorious liberty of the children of God. For we know that the whole **creation** groans and labors with birth pangs together until now."

Colossians 1:15—"He is the image of the invisible God, the firstborn over all **creation**."

1 Peter 4:19—"Therefore let those who suffer according to the will of God commit their souls *to Him* in doing good, as to a faithful **Creator**."

2 Peter 3:4—"and saying, "Where is the promise of His coming? For since the fathers fell asleep, all things continue as *they were* from the beginning of **creation.**"

According to *An Expository Dictionary of New Testament Words*, the Greek verb *ktizō* "signifies, in Scripture, 'to create,' always of the act of God, whether (*a*) in the natural creation, Mark 13:19; Rom. 1:25 (where the title 'The Creator' translates the article with the aorist participle of the verb; ... or (*b*) in the spiritual creation, Eph. 2:10, 15; 4:24; Col. 3:10" (p. 137; published as a component of *Vine's Expository Dictionary of Biblical Words*). The noun form *ktisis* signifies "primarily 'the act of creating,' or 'the creative act in process'" (p. 137). Clearly, then, the New Testament affirms the creative acts of God, contrary to the naturalistic, materialistic evolutionary theory.

Other Passages Affirming Divine Creation

Many other passages support supernatural creation: The following examples are representative of Old and New Testament teaching in this regard:

Genesis 1:7—"Thus God **made** the firmament, and divided the waters which *were* under the firmament from the waters which *were* above the firmament; and it was so."

Genesis 1:16—"Then God **made** two great lights: the greater light to rule the day, and the lesser light to rule the night. He **made** the stars also."

Genesis 1:25—"And God **made** the beast of the earth according to its kind, cattle according to its kind, and everything that creeps on the earth according to its kind. And God saw that *it was* good."

Genesis 2:9—"And out of the ground the Lord God **made** every tree grow that is pleasant to the sight and good for food. The tree of life *was* also in the midst of the garden, and the tree of the knowledge of good and evil."

Genesis 2:22—"Then the rib which the Lord God had taken from man He **made** into a woman, and He brought her to the man."

Genesis 7:4—"For after seven more days I will cause it to rain on the earth forty days and forty nights, and I will destroy from the face of the earth all living things that I have **made**."

Genesis 9:6—"Whoever sheds man's blood, By man his blood shall be shed; For in the image of God He **made** man."

God's Creative Acts Took Six Days

Since some people contend that although God created the universe, He did it by a gradual process of evolution over a long period of time, it is important to establish from the Scriptures that the time period of creation was six days, not millions or billions of years. A partial listing of Bible passages will sufficiently demonstrate this truth.

Genesis 1:31—"Then God saw everything that He had made, and indeed *it was* very good. So the evening and the morning were the **sixth day**." That the days of creation were 24-hour days rather than long periods of time is indicated by the repeated reference to "the evening and the morning" (Genesis 1:5, 8, 13, 19, 23, 31).

Exodus 20:11—"For *in* **six days** the Lord made the heavens and the earth, the sea, and all that *is* in them, and rested the seventh day...."

Exodus 31:17—"It *is* a sign between Me and the children of Israel forever; for *in* six days the Lord made the heavens and the earth, and on the seventh day He rested and was refreshed.'"

Matthew 19:4—"And He answered and said to them, 'Have you not read that He who made *them* **at the beginning** 'made them male and female.'" Observe that Jesus referred to "the beginning" as recorded in Genesis 1, in which God's creative acts involved six days. If the days of Genesis 1 were long periods of time (e.g., millions or billions of years), the creation of man and woman on the *sixth day* would not have been "at the beginning" but a long time later.

God's Creative Acts Were Immediate

Furthermore, the account in Genesis 1 indicates the *immediate* implementation of God's creative commands, not a gradual process requiring millions or billions of years:

Genesis 1:3—"Then God said, 'Let there be light'; **and there was light.**"

Genesis 1:9—"Then God said, 'Let the waters under the heavens be gathered together into one place, and let the dry *land* appear'; **and it was so.**"

Genesis 1:11—"Then God said, 'Let the earth bring forth grass, the herb *that* yields seed, *and* the fruit tree *that* yields fruit according to its kind, whose seed *is* in itself, on the earth'; **and it was so.**"

Genesis 1:14-15—"Then God said, 'Let there be lights in the firmament of the heavens to divide the day from the night; and let them be for signs and seasons, and for days and years;

and let them be for lights in the firmament of the heavens to give light on the earth'; **and it was so.**"

Genesis 1:24—"Then God said, 'Let the earth bring forth the living creature according to its kind: cattle and creeping thing and beast of the earth, *each* according to its kind'; **and it was so.**"

Commenting on such statements, Riegle (1971) wrote: "The Bible story implies that the acts of the Creation were **instantly accomplished**" (p. 84, emphasis added). Matthew Henry (1828) stated that "the light was made by the word of God's power. He said, *Let there be light*; he willed and appointed it, and it was done **immediately**: *there was light*" (p. 3; emphasis added).

Biblical Evidence Against Human Evolution

Henry Morris (2002) outlined seven lines of Biblical evidence against human evolution, emphasizing "the unbridgeable gap between man and the animals" (p. 364). His points are summarized as follows:

1. *Man's dominion....* (Gen. 1:28). The animals were all created to serve man, not to compete with him in an evolutionary struggle for survival.

2. *Man's body specially formed....* Adam and Eve were individually formed directly by God himself....

3. *No help-meet among the animals.* When Adam was instructed to name the animals ... there were none that were sufficiently like him to be a "helper fit for" him. This indicates there were none whose immediate past ancestry he shared.

4. *Adam's return to the dust....* The "dust of the ground" from which he had been formed (Gen. 2:7) could not have been a long evolutionary development, as theistic evolutionists had alleged.

5. *Eve's unique creation.* It is impossible to explain the special formation of Eve's body out of Adam's side by the Lord in terms of any kind of evolutionary development from an animal ancestry.

6. *Chronology of man's creation.* According to the testimony of Christ himself ... (Mark 10:6, quoting Gen. 1:27) ... man and woman were made, not after four billion years of evolutionary development, but from the beginning of the creation.

7. *Distinctiveness of human flesh....* "All flesh is not the same flesh: but there is one kind of flesh of men, another flesh of beasts, another of fishes, and another of birds" (1 Cor. 15:39). (pp. 364-365)

Although members of our society are bombarded with evolutionary propaganda through media, amusement parks, school classrooms, and almost every segment of our cultural setting, and Christians are often misled into accepting the assumptions of naturalistic evolution, the Bible makes it abundantly clear that man did not evolve from lower forms of life. God created human beings in His image.

God's Power Demonstrated

In addition to the obvious rejection of anything miraculous by atheists and agnostics, some individuals who believe in some type of God (e.g., Deists) reject the miracles recorded in the Bible. Consequently, such individuals generally espouse some form of evolutionary view of origins rather than special creation as described in Genesis 1-2. In many cases this position involves skepticism regarding any supernatural intervention in the origin and development of things.

Clearly, the Bible refers often to miraculous events beginning with God's supernatural creation: "Then God said ... and it was so" (cf. Genesis 1:6-7, 9, 11, 15, 24, 30). The belief system of some individuals predisposes them to reject all references to the miraculous. However, for those who accept the evidence for even one miracle (e.g., the resurrection of Jesus from the dead), it would be inconsistent to reject the other miracles recorded in the Bible, including the miracles of creation in Genesis 1.

Summary

Although the scriptural citations have not been exhaustive, sufficiently many have been provided to demonstrate overwhelmingly that the Bible (in both Old and New Testaments) declares that God is the Creator and that all things were made by Him miraculously during the six days of the creation week. Rejecting these numerous statements from God's revelation and contending that all things have come to exist as a result of billions of years of naturalistic processes is untenable for the true Bible believer.

Nevertheless, some choose to reject the Biblical explanation of origins. For example, Montagu (1984) wrote: "Whatever the historical antecedents of Genesis, it represents but one of innumerable creation myths which different people at different times have invented in order to account for the manner in which Earth and everything upon it came into being" (p. 6). In view of such rejections of Bible teaching by many people, the next chapter will address scientific reasons for doubting evolutionary approaches to the origin and development of all things.

CHAPTER 3

ARGUMENTS BASED ON SCIENCE

Since the teaching of Scripture will be unconvincing to many people, this chapter addresses numerous scientific principles and ideas that undermine or call into question the belief that naturalistic processes operating over billions of years adequately explain the existence of all things today.

Errors in Science

Although many people consider scientific claims to be

dependable and absolute, there have been numerous instances of erroneous or fraudulent claims by scientists. These instances, which are not intended to exhaust all of the available evidence, provide words of caution about blindly accepting the claims made on behalf of naturalistic evolution and its various lines of evidence. A few illustrative instances will be cited to establish the point that scientific theories and claims have sometimes been wrong. Consequently, it is reasonable to be suspicious of claims made even by scientists, especially when those claims are not supported by evidence but involve speculation, as frequently indicated by words such as "probably," "maybe," "we believe," or similar terms.

Haeckel's Biogenetic Law. Hotton (1968) cites the case of Haeckel's *Biogenetic Law*: "Among Darwin's champions was the German biologist Ernst Haeckel, who is known for the now-discarded 'biogenetic law' that in the embryo an organism passes through a series of stages repeating the evolutionary history of its ancestry" (p. 16). However, it was discovered that he "fudged" his drawings that allegedly supported his ideas.

Feldkamp (2002) wrote: "The German zoologist Ernst Haeckel (1834-1919) ... declared that 'ontogeny recapitulates phylogeny.' This statement can be translated to 'embryological development repeats evolutionary history.' We now know that this is a bit of an exaggeration" (pp. 290-291). That is an understatement! Haeckel's drawings were intentionally doctored to give the appearance of greater similarity among various living creatures during their embryonic development. Nevertheless, the same basic idea is believed and taught today by some evolutionists.

Human evolution. Human evolution is another of the areas wherein scientists have changed their opinions over time. While it is honorable to be willing to change one's opinions when additional evidence is discovered, it nevertheless points to the very tentative nature of "scientific" beliefs. For example, scientists once believed that humans evolved much more recently than now thought and that they originated in different parts of the world:

"Before 1925, many scientists believed that the first humans had evolved in Europe or Asia no more than 200,000 years ago" (Feldkamp, 2002, p. 322).

Feldkamp (2002) further describes how evolutionists' views of human evolution have changed:

> This find [*Australopithecus afarensis*] changed many ideas about the evolution of humans. It was generally thought that the hallmarks of hominids—bipedalism and a large brain with areas dedicated to higher reasoning and the production of speech—had all evolved at the same time. But the new fossil showed that upright walking had apparently come before many of these other adaptations that make the hominids unique among the anthropoid primates. (p. 324)

Mader (2010) also admitted that scientists had been mistaken in their earlier opinions regarding human evolution. The fact that chimpanzees have 48 chromosomes and humans have 46 was considered so significant that this evidence prompted scientists to classify the great apes and humans into different families. Through the studies at Yale University in 1991, that opinion changed (p. 242).

Among the most outstanding examples of the speculations scientists have proposed, often on the basis of very little evidence, is the famous case of Nebraska Man. In 1917 Harold Cook found a tooth in Upper Snake Creek beds of Nebraska. The find was named *Hesperopithecus haroldcookii*, in recognition of its discoverer. *Hesperopithecus* means "ape of the western world," and the find was used in support of the theory of evolution in the famous Scopes Monkey Trial in 1925.

After several scientists proclaimed that it was a specimen of a manlike ape, an illustration appeared in the *Illustrated London News*. Henry Osborn, who received the tooth in March 1922, said the illustration of Nebraska Man and his habitat was "a figment of the imagination of no scientific value, and undoubtedly inaccurate" (retrieved from https://en.wikipedia.org/wiki/Nebraska_Man, October 4, 2015). However, although not a deliberate deception, the classification was discovered to be a mistake when other parts of the skeleton were unearthed. The tooth was actually from an extinct peccary (*Prosthennops serus*). Nevertheless, Nebraska

Man was given as an example of a "missing link" in the ancestry of man as late as 1943, many years after the error was exposed (Wysong, 1976, pp. 295-296).

Piltdown Man was also "claimed as a crucial and significant find proving man's evolution.... [However, it] was later exposed as a reconstruction based upon an ape's jaw and human skull fragments that had been fraudulently doctored to give a 'genuine appearance'" (Wysong, 1976, p. 296). However, it was still cited in textbooks as evidence for human evolution almost 40 years after it had been demonstrated to be fraudulent" (p. 296).

Southwest Colorado Man was believed by some scientists to link men and primates. However, the single tooth that served as the basis for this belief was discovered "to belong to an extinct Eocene horse" (Wysong, 1976, p. 296).

Numerous other mistakes and frauds have occurred in a vain effort by scientists to establish the evolutionary origins of human beings. Attempts are often made to excuse these mistakes and frauds relating to evolution; however, that mistakes were made by some scientists is indisputable, often because they reached conclusions based upon faulty or incomplete evidence or preconceived ideas. Sometimes scientists were guilty of fabricating or altering evidence to support a theory they advanced. Instances of fraud are always inexcusable. These occurrences demonstrate the need for us to be cautious, and sometimes suspicious, of "scientific" claims. This is certainly the case with evolutionary claims.

Lamarckism Inheritance of Acquired Characteristics. Chevalier de Lamarck proposed a theory that living things inherit acquired characteristics. Mader (2010) observed that Lamarck sought "to explain the process of adaptation to the environment [by proposing] ... the idea of *inheritance of acquired characteristics*" (p. 250).

Hardin (2010) notes the great appeal of Lamarck's ideas, stating:

> Perhaps no doctrine in biology in the last two hundred years has had so great an appeal and has necessitated so many experiments with negative conclusions as has the Lamarckian doctrine of the inheritance of acquired characteristics.... let it be flatly stated that, with few and quite unimportant qualifications, there's nothing in Lamarckism. Acquired characteristics are not inherited. (p. 119)

Feldkamp (2002) also acknowledges the error of Lamarck's theory:

> The French scientist Jean Baptiste de Lamarck (1744-1829) was one of the first to propose a unifying hypothesis of species modification.... He hypothesized that acquired traits were passed on to offspring.... Lamarck's hypotheses were fiercely attacked.... Although Lamarck's hypothesis ... was easily disproved, his work was an important forerunner of modern evolutionary theory. (p. 283)

Lamarck's hypothesis, the inheritance of acquired characteristics, has been clearly demonstrated to be false. It stands as a prime example of the fact that scientists have sometimes held positions that were actually wrong. On that basis, especially in view of the many weaknesses cited for evolution, it is completely unreasonable for some scientists to claim that evolution is a fact.

Spontaneous Generation. In order for the naturalistic view to establish a beginning point it is necessary that something nonliving became living at some stage in the process. Consequently, despite the lack of evidence for such a view, the concept of spontaneous generation was held in the past. Hotton (1968) comments, "The notion that under certain circumstances life could arise from nonliving matter ... (it became known as the theory of 'spontaneous generation') obviously resulted from insufficient attention to external phenomena ... the theory was generally discredited well before Darwin's time" (p. 8).

Smallwood and Green (1971) explain that "reputable scientists, who otherwise made significant contributions to man's knowledge, were not immune to some of the fanciful superstitions associated with the theory. In the 1600's, for instance, the noted Belgian scientist Jan van Helmont ... devised a recipe for generating mice from a shirt!" (p. 229).

Despite the scientific evidence against it, this theory of spontaneous generation persisted among some scientists until

1864. Smallwood and Green (1971) state: "As recently as 1864 some scientists believed that organisms could arise spontaneously from nonliving matter. This was the so-called theory of spontaneous generation…. Pasteur's inquiry, which proved that an organism could only arise from another organism, helped focus attention upon the cell" (p. 37).

Commenting further on Pasteur's experiments, Smallwood and Green say, "Louis Pasteur in 1864 convincingly disproved the theory of spontaneous generation. He proved conclusively that micro-organisms are no different from any other form of life; they can only be produced from others of their own kind" (p. 231).

Despite the fact that all scientific experimentation, as well as every observation throughout all of history, confirms that spontaneous generation of living things from non-living things does not occur, evolutionists are left with no alternative to the assertion that it has occurred at some point in the remote past. For example, Simpson wrote that "virtually all biochemists agree that life on earth arose spontaneously from non-living matter" (p. 771).

David White, in the first cross-examination period of a December 1993 debate on evolution, when asked if spontaneous generation had ever occurred, asserted, "That's what the evidence shows." However, he offered no evidence in support of this answer.

Even researchers sometimes find the claim that spontaneous generation has occurred to be unsatisfactory. For example, Gurin (1981) wrote: "Researchers interested in life's origin find such traditional explanations far from satisfactory.... No one really understands how these components could have spontaneously assembled to form a living cell ... it is almost impossible to conceive of such a system arising spontaneously and by chance ..." (p. 17).

Otto, Towle, and Bradley (1981) describe the replacement of belief in spontaneous generation with the concept of biogenesis: "The theory of spontaneous generation, accepted for centuries without any real evidence, was defeated. From Pasteur's time to the present, all experiments have supported biogenesis: Life comes only from life" (p. 25). Consequently, despite the overwhelming scientific evidence against the occurrence of spontaneous generation, evolutionists believe that it occurred. The evolutionary process cannot even begin without some nonliving thing changing to a living thing at some point in time. The idea that such an event took place billions of years ago does not satisfy the demands of evidence.

"Lucky Monsters." Because the abundant fossil record does not support the gradual transformation required by the usual Darwinian theory of evolution, some scientists proposed more rapid changes. Hotton (1968) states:

> The discovery of mutations suggested that evolution might take place in a series of large jumps. As 'lucky monsters' appeared that were fortuitously adapted to a different set of conditions than their parents.... These ideas were soon abandoned, however, as it became clear that large mutations are in general deleterious, and that the odds against

mutations producing a viable form significantly different from its parents are quite high. (p. 18)

Although some express disdain for the term "lucky monsters," several prominent evolutionists have espoused the "punctuated equilibrium" idea. For example, Morris (2002) refers to the ideas of Stephen Jay Gould and Niles Eldredge. In a statement by Gould from an article in *Paleobiology* (6:1, p. 125), he writes: "Thus, our model of 'punctuated equilibria' holds that evolution is concentrated in events of speciation and that successful speciation is an infrequent event punctuating the stasis of large populations that do not alter in fundamental ways during the millions of years that they endure" (Morris, 2002, p. 314). They claim that theirs is "a better model than the 'slow-and-gradual' concept of the neo-Darwinists" (p. 314). If the gradualistic view had been supported by sufficient evidence, paleontologists would not have needed to develop the hypothesis of punctuated equilibrium. The fact that scientists set forth this hypothesis points to the weakness of the fossil record in supporting the Darwinian model.

Dating Methods

The theory of evolution requires an incredible amount of time for its alleged gradual processes to occur. It is noteworthy that extremely long periods of time alone will not establish the truth of evolution; however, without billions of years of time, evolution as believed and taught today is impossible. Consequently, efforts have been made to establish incredibly long periods of time during which evolution is at least arguable.

Although the Carbon-14 dating method applies to relatively short geologic time periods (thousands rather than millions or billions of years), the assumptions upon which this method is based are similar to those of other long-term radiometric dating methods. Antevs (1957) commented:

> In appraising C-14 dates, it is essential always to discriminate between the C-14 age and the actual age of the sample. The laboratory analysis determines only the amount of radiocarbon present…. However, the laboratory analysis does not determine whether the radiocarbon is all original or is in part secondary, intrusive, or whether the amount has been altered in still other irregular ways besides by natural decay.

(p. 129)

Referring to the uncertainties, and sometimes absurdities of radiometric dating, Reed (1959) remarked, "Although it was hailed as the answer to the prehistorian's prayer when it was first announced, there has been increasing disillusion with the method because of the chronological uncertainties, in some cases absurdities, that would follow a strict adherence to published C-14 dates" (p. 1630).

Numerous examples of ridiculous results of radiometric dating have been obtained. For example, living snails have been dated as 2,300 years old by the C-14 method (Keith and Anderson, 1963, p. 634). Wysong (1976) cites several instances in which absurd results have been obtained by using C-14 dating methods, including:

New wood from actively growing trees has been dated ... at 10,000 years!

Mortar from the Oxford Castle in England was assigned an

age ... of 7,370 years, but the castle was built only 785 years ago.

Freshly killed seals have been dated at 1,300 years, and mummified seals dead no longer than 30 years have been dated up to 4,600 years. (p. 151)

Similarly ridiculous dates have been assigned to objects dated by other methods (e.g., potassium-argon, uranium-lead, and thorium-lead). This is associated with the fact that similar assumptions serve as the basis for the various dating methods.

One of the common assumptions of these dating methods is that the decay rate is constant; it is unaffected by external influences. However, "alterations in decay rate have been measured in modern laboratories. There is also evidence that decay rate has varied in the past" (Wysong, 1976, p. 152). He concludes: "If decay rate can be changed by bombarding nuclei with high energy particles, if decay rate can change due to chemical combinations and physical pressure, and if there is halo evidence that decay rate has changed over geological time, then the assumption that decay rate ... has not changed ... is truly an assumption" (p. 153).

In view of the questionable assumptions of the various dating methods commonly used by evolutionists to indicate vast geologic time, evolutionary conclusions related to these time measurements are suspect.

Arguments Based On Supposition

It is quite common to read material that teaches naturalistic evolution in which the authors repeatedly use words to indicate that they argue based on assumptions and suppositions. For example, in attempting to explain the origin of flight one text states:

> The evolution of a flying animal from nonflying ancestors entails many changes in anatomy, physiology, and behavior. **According to one hypothesis**, the ancestors of birds were tree dwellers that ran along branches and occasionally jumped between branches and trees. Wings that allowed these animals to glide from tree to tree evolved. Once gliding was possible, the ability to fly by flapping the wings evolved. **Another hypothesis** draws on the fact that dinosaurs most closely related to birds were terrestrial and states that the

evolution of birds must have occurred on the ground, not in the trees. Wings **may** have originally served to stabilize the animals as they leapt after prey. Or they **may** have been used for trapping or knocking down insect prey. Over generations, the wings became large enough to allow the animal to become airborne. (pp. 862-863; emphasis added)

Hotton (1968) admits the use of speculation by scientists: "Two other lines of evidence, from geology and biochemistry, permit **speculation** about a possible origin of life from nonliving matter" (p. 19; emphasis added).

Scientists often base their notions upon assumptions rather than facts. For example, Mader (2010) states: "Because dinosaurs are classified as reptiles, paleontologists at first **assumed** that dinosaurs behaved in the same manner as today's reptiles" (p. 290; emphasis added). She later says that "the dinosaurs **may** have benefitted from the mass extinction at the end of the Triassic…" (p. 295; emphasis added). Discussing the supposed evolution of the horse, she comments, "Fossils named *Hyracotherium* have been designated as the first **probable** members of the horse family" (p. 286; emphasis added).

Discussing the mass extinctions of dinosaurs but not mammals during the Mesozoic era, Mader speculates: "**Perhaps** their small size and lack of specialization helped them survive" (p. 29; emphasis added). The use of these and other words indicates that much of evolutionary theory consists of speculation rather than facts. On numerous occasions scientists who have subscribed to a particular idea have changed their minds because additional evidence came to light.

Another example in which scientific opinion shifted concerns Charles Lyell's view of uniformitarianism. Mader (2010) describes this change in scientific opinion: "This idea of slow geologic change is still accepted today, although modern geologists have concluded that rates of change have not always been uniform" (p. 251). Previously, scientists believed that geologic changes occurred at a uniform rate, which was the basis for the term *uniformitarianism*.

In addition, scientists have often formed hypotheses without sufficient evidence to warrant their beliefs: "Scientists had **always thought** whales had terrestrial ancestors. Now, fossils have been discovered that support this hypothesis" (Mader, 2010, p. 256; emphasis added). Observe that scientists formed the opinion first, but evidence that they claim supports the view was found later! Changing one's conclusion on the basis of newly discovered evidence is commendable, but forming opinions without sufficient evidence leads to erroneous positions. In the case of naturalistic evolution in all its aspects, numerous opinions have been propagated simply because divine creation was rejected as an acceptable option.

In *Teaching About Evolution and the Nature of Science*, the National Academy of Sciences (1998) asserted: "A particularly interesting example of contemporary evolution involves the 13 species of finches studied by Darwin on the Galapagos Islands, now known as Darwin's finches" (p. 19). The authors seem to ignore the obvious fact that the "finches" were still "finches"; there is no proof whatsoever that some other creature evolved over a long period of time into finches, or that finches evolved over a long period of time to some other creature. However, this type of macroevolution is demanded by the theory of evolution. The variation observed among Darwin's finches is within the range of variation designed by God, just as there is considerable variation among different breeds of dogs, which are all of the same species, and among human beings, who are all the same species.

The famous case of light- and dark-colored peppered moths in Great Britain has often been cited as an example of evolution. However, neither type evolved into the other, nor did either type evolve from or into something else; their proportions simply changed due to industrialization (Shute, 1961, p. 103; Wysong, 1976, p. 313). Evolutionists seem oblivious to these simple facts, and they continue pretending that peppered moth prove evolution.

Biogenesis

Otto, Towle, and Bradley (1981) point out the truth of the principle of biogenesis: "From Pasteur's time to the present, all experiments have supported biogenesis: Life comes only from life" (p. 25). They observe that at one time people "believed that nonliving things would produce living organisms. This theory has been replaced by a belief in biogenesis, which means all life comes from life" (p. 25).

Smallwood and Green (1971) discuss the experiments of Francesco Redi, who "strengthened the case against the theory of spontaneous generation. This alternate theory that Redi espoused held that all life originates from life. It was called *biogenesis*." (p. 230). Wysong (1976) observes that "the law of biogenesis finds its starting foundation in the work of Francesco Redi in the 1600's. Since that time the hypothesis of biogenesis has known no exceptions and proves perfectly predictable" (p. 185).

Every observation by humans throughout history supports

the conclusion that a living thing always comes from a living thing. Nevertheless, God has incorporated into each living thing the capability for limited variation. This is obvious when we observe the appearances of human beings, even within the same family. Although each individual is human, differences in many characteristics can be seen. The same is true of other types of living things. For example, Smallwood and Green (1971) state: "All domestic dogs belong to one species. Yet, after looking at all their various sizes, shapes, and patterns of behavior, it is not at all obvious that all dogs are of 'one kind'" (p. 87).

Lipson (1980) questioned the Darwinists' explanations for the origin of living things. After reviewing many problems associated with obtaining living things from nonliving, he asked, "If living matter is not, then, caused by an interplay of atoms, natural forces, and radiation, how has it come into being?" (p. 138). Dismissing any type of directed evolution, he concluded, "I think, however, that we must go further than this and admit that the only acceptable explanation is **creation**" (p. 138; emphasis in original). He then remarked, "I know that this is anathema to physicists, as indeed it is to me, but we must not reject a theory we do not like if the experimental evidence supports it" (p. 138).

Virchow's Principle

In 1858 the experiments of Rudolf Virchow demonstrated that "cells do not arise from amorphous exudate, but rather from preexisting cells, and the labors of countless scientists in all of the various disciplines of biology since, have established the law of biogenesis—life springs from preexisting life" (Wysong, 1976, pp. 180, 182). Ackerknecht (1973) comments concerning Virchow's conclusions: "His aphorism 'omnis cellula e cellula' (every cell arises from a pre-existing cell) ranks with Pasteur's 'omne vivum e vivo' (every living thing arises from a preexisting living thing) among the most revolutionary generalizations of biology" (p. 35).

Smallwood and Green (1971) express complete agreement with their peers: "Where a cell arises, there a cell must have been before, even as an animal can come from nothing but an animal, a plant from nothing but a plant. Thus in the whole series of living things there rules an eternal law of continuous development. There is no discontinuity nor can any developed issue be traced back to anything but a cell" (p. 38).

Naturalistic evolution's biopoiesis or abiogenesis, although transferred into the distant past, contradict Virchow's Principle and the results of experimentation by scientists such as Redi, Spallanzani, and Pasteur.

Origin of Life Experiments

H. P. Yockey (1977) describes the faith-based views of some scientists regarding the origin of life:

> Research on the origin of life seems to be unique in that the conclusion has already been authoritatively accepted.... What remains to be done is to find the scenarios which describe the detailed mechanisms and processes by which this happened. One must conclude that, contrary to the established and current wisdom, a scenario describing the genesis of life on earth by chance and natural causes which can be accepted on the basis of fact and not faith has not yet been written. (p. 377)

H. S. Lipson (1980), a respected British physicist and

evolutionist, commented on his longstanding interest in the origin of life and noted that "evolution became in a sense a scientific religion; almost all scientists have accepted it and many are prepared to 'bend' their observations to fit with it" (p. 138). Lipson then pondered how successfully evolution had withstood scientific testing, concluding:

> I have always been slightly suspicious of the theory of evolution because of its ability to account for any property of living beings. I have therefore tried to see whether biological discoveries over the last thirty years or so fit in with Darwin's theory. I do not think that they do. To my mind, the theory does not stand up at all. (p. 138)

Patterson (2007) describes the absence of proof for the naturalistic origin of life on earth:

> No known laws of nature allow complex, living, information-containing systems to develop from the random interactions of matter. Yet, this is what is required in order for life to have evolved in the universe. Creationists accept that life appeared on earth as a direct creative act of God. The design and information that we see in all living things is a result of an intelligence—not random occurrences. Many biologists try to separate the origin of life ... and the universe from discussions of biology. These scientists recognize the challenge presented by chemical evolution, but their commitment to naturalism

leaves them no other choice. (p. 134)

Several scientists have attempted to develop experiments that shed light on how life can originate from nonliving materials. Smallwood and Green (1971) describe the approach of Oparin, one of the pioneers in biopoiesis:

> One of the leading hypotheses ... is the *Oparin hypothesis*, named after its principal author, A. I. Oparin, a noted Russian biochemist.... Oparin outlined a sequence of events that *could* have led to the origin and evolution of life on the earth. Oparin's hypothesis is based on the set of conditions that he assumed prevailed on the earth at the time these events were occurring.... This is a controversial matter among scientists, since no single set of primitive atmospheric conditions has been agreed upon by all investigators." (pp. 231-232)

The Miller-Urey model is the most widely acclaimed origin of life experiment. Among the defects of this model is the assumption of "a reducing atmosphere in the early earth.... [Urey] argued that the atmosphere would have contained abundant amounts of ammonia, methane, hydrogen and water vapor" (Wysong, 1976, p. 220). Wysong lists some frequently offered objections to the legitimacy of the Miller-Urey model:

A. No evidence forces the use of this concentration [of methane and ammonia to approximate the constituents of glycine].... If the concentrations of methane and ammonia were high, no amino acids would obtain.

B. A methane-ammonia reducing atmosphere would be highly toxic to life.

C. The geochemist, Abelson, states: "The hypothesis of an early methane- ammonia atmosphere is found to be without solid foundation and indeed is contradicted." ... [Abelson concludes:] "What is the evidence for a primitive methane-ammonia atmosphere on earth? The answer is that there is no evidence for it, but much against it." ...

D. The Miller apparatus is designed to produce.... [Hull observes that "the mere fact that a chemist can carry out an organic synthesis in the laboratory does not prove that the same synthesis will occur in the atmosphere or open sea without the chemist."

E. The experiments have provided only the building blocks of life, never the highly ordered, information carrying, optically active (D or L) molecules ... of life—let alone anything even suggestive of a living entity. (p. 222)

What should be apparent to all reasonable observers as an obvious defect of origin of life experiments is that the results are not produced by purely naturalistic occurrences over long periods of time. Behind each experiment is careful planning and deliberate design by an individual or group of individuals. They design and set up the apparatus, procure and measure the chemicals to be used, etc. Therefore, these experiments are not indicative of the results of chance processes occurring over millions or billions of years.

The Fossil Record

Proponents of evolution have frequently argued that the fossil record proves that evolution has occurred: "The best evidence for evolution comes from fossils, the physical remains of organisms that lived on Earth between 10,000 and billions of years ago…. The fossil record is the history of life recorded by fossils and the most direct evidence we have that evolution has occurred" (Mader, 2010, p. 255).

However, for that to be so, numerous transitional forms should have been discovered. Charles Darwin wrote:

> As by this theory, innumerable transitional forms must have existed. Why do we not find them imbedded in the crust of the earth? Why is all nature not in confusion instead of being as we see them, well-defined species? Geological research

does not yield the infinitely many fine gradations between past and present species required by the theory; and this is the most obvious of the many objections which may be argued against it. The explanation lies, however, in the extreme imperfection of the geological record. (p. 49)

Clark (1930) stated: "No matter how far back we go in the fossil record of previous animal life upon the earth, we find no trace of any animal forms which are intermediate between the various major groups or phyla" (p. 189). If gradual evolution over a long period were so, the fossil record should give some indication by preserving evidence of numerous transitional forms. But Clark observes, "Since we have not the slightest evidence, either among the living or the fossil animals, of any intermediate types following the major groups, it is a fair supposition that there never have been any such intergrading types" (p. 196).

Mader (2010) admits that the fossil record is not as definitive as commonly supposed: "One of the advantages of fossils is that they can be dated, but unfortunately it is not always possible to tell to which group, living or extinct, a fossil is related.

For example, at present, paleontologists are discussing whether fossil turtles indicate that turtles are distantly or closely related to crocodiles" (p. 303).

Stahl (1974) highlights the inadequacy of the fossil record concerning the supposed evolution of mammals: "Because of the nature of the fossil evidence, paleontologists have been forced to reconstruct the first two-thirds of mammalian history in great part on the basis of tooth morphology" (p. 401).

Watson (1982) describes the inadequacy of the fossil record in demonstrating the evolution of apes: "Modern apes, for instance, seem to have sprung out of nowhere. They have no yesterday, no fossil record. And the true origin of modern humans—of upright, naked, tool-making, big-brained beings—is, if we are to be honest with ourselves, an equally mysterious matter" (p. 44).

Regarding the failure of the fossil record to provide the many transitional forms required by Darwinian concepts, even after well over a century of searching, Heribert-Nilsson (1982) observed that "it is not even possible to make a caricature of evolution out of paleo-biological facts. The fossil material is now so complete that the lack of transitional series cannot be explained by the scarcity of the material. The deficiencies are real, they will never be filled" (p. 1212). Neville (1960) concludes: "There is no need to apologize any longer for the poverty of the fossil record. In some ways it has become almost unmanageably rich, and discovery is outpacing integration…. The fossil record nevertheless continues to be composed mostly of gaps." (pp. 1, 3).

Simpson (1953) had made a similar determination: "It remains true, as every paleontologist knows, that most new species, genera, and families and that nearly all new categories above the level of families appear in the record suddenly and are not led up to by known, gradual, completely continuous transitional sequences" (p. 360).

Corner (1961) expresses the opinion regarding the fossil evidence provided by plants: "I still think that, to the unprejudiced, the fossil record of plants is in favour of special creation." (p. 97). Riegle (1971) similarly comments:

> Many fossils have been found, but none show the very important transitional stages between the various types of animals. If animals really evolved from
>
> other animals which were not as far advanced, and if these transitional stages really took millions of years, it seems that there should exist fossils of some
>
> 'in-between' forms which would show that these changes did in fact take place. This is one of the greatest weaknesses in the story of evolution." (p. 86)

Raup (1979) acknowledges the inadequacy of the fossil evidence: "Darwin's theory of natural selection has always been closely linked to evidence from fossils, and probably most people assume that fossils provide a very important part of the general argument that is made in favor of Darwinian interpretations of the history of life. Unfortunately, this is not strictly true" (p. 22).

Kelso (1974) admitted that "the transition from insectivore to primate is not documented by fossils. The basis of knowledge about the transition is by inference from living forms" (p. 142).

Thompson (1983) concludes, "Rather than supporting evolution, the breaks in the known fossil record support the creation of major groups with the possibility of some limited variation within each group" (p. 76).

Raup (1979) reached the same conclusion:

> Instead of finding the gradual unfolding of life, what geologists of Darwin's time, and geologists of the present day, actually find is a highly uneven and jerky record; that is, species appear in the sequence very suddenly, show little or no change during their existence in the record, then abruptly go out of the record. And it is not always clear, in fact it's rarely clear, that the descendants were actually better adapted than their predecessors. In other words, biological improvement is hard to find. (p. 25)

Gould (1977) observed that natural groups of fossil organisms appeared abruptly and that systematic gaps existed between different groups of fossil organisms. He commented, "New species almost always appeared suddenly in the fossil record with no intermediate links to ancestors in older rocks of the same region" (p. 12).

The history of most fossil species includes two features particularly inconsistent with gradualism:

> 1. Stasis....
>
> 2. Sudden appearance. In any local area, a species does not arise gradually by the steady transformation of its ancestors; it appears all at once and "fully formed." (p. 13)

Raup (1979) describes the shortcomings of Darwinian ideas as demonstrated in the fossil record:

> We are now about 120 years after Darwin and the knowledge of the fossil record has been greatly expanded. We now have a quarter of a million fossil species but the situation hasn't changed much. The record of evolution is still surprisingly jerky and, ironically, we have even fewer examples of evolutionary transition than we had in Darwin's time. By this I mean that some of the classic cases of Darwinian change in the fossil record, such as the evolution of the horse in North America, have had to be discarded or modified as a result of more detailed information.... So Darwin's problem has not been alleviated in the last 120 years.... (p. 25)

Mutations

The theory of evolution as commonly represented requires the transmission of traits resulting from genetic mutations. However, the scientific evidence is contrary to the accumulation of favorable traits by this means. Smallwood and Green (1971) observe that "experimental evidence indicates that mutations occur quite often but that most of them are either harmful or 'neutral.' Mutations in this latter group would not confer any favorable characteristics upon the individual possessing them. Favorable mutations are relatively rare events" (p. 221).

Vestigial Structures

"Vestigial structures occur because organisms inherit their anatomy from their ancestors; they are traces of an organism's evolutionary history" (Mader, 2010, p. 257). Actually, the number of alleged vestigial structures has decreased as knowledge has increased. Some of those body parts that were once thought to have no function are now known to contribute in important ways to the functioning of the human body. The argument for evolution based upon vestigial structures is an argument from ignorance. It would be much more accurate for scientists to acknowledge that they just do not know the function(s) of so-called vestigial structures, rather than claiming that they have no function and use them as evidence for gradualistic evolution.

The Physical Universe

In addition to the inadequacy of evolutionary theories, particularly the dominant Darwinian perspective, to account for the existence of the vastly complex realm of living things, evolutionists are also unable to explain numerous observable phenomena in the non-living realm. Their efforts usually consist of unsupported speculations, as the frequent use of such words as "possibly," "maybe," "might," and "some scientists think" indicate. Although an exhaustive compilation of such statements will not be attempted, a few examples are provided as illustrative of this approach.

Among the examples of the absence of adequate explanations relating to observations of the non-living realm are the following unanswered questions raised by Riegle (1971): "Why do the planets contain many elements different from those believed to be found on the sun? ... How did it happen that all the satellites of a single planet do not revolve about the parent planet in the same direction?" (pp. 82-83). He argues that "no theory can explain why three of Jupiter's twelve moons, one of Saturn's nine, and that of Neptune revolve in a direction opposite that of their parent planets" (p. 17). Naturalistic explanations are inadequate. It is simple (and accurate) to explain any such observations as a result of God's creative acts that display His mighty power.

Probability

Numerous mathematical calculations have been presented to demonstrate the extreme unlikelihood of the various components of the theory of evolution. "When considering the probability of the assembly of a DNA molecule," the chance is so remote that "the probability is clearly outside the realm of [reasonable] possibility" (Patterson, 2007, p. 147). "Harold J. Morowitz, professor of biophysics at Yale, has calculated that the formation of one *E. coli* bacteria in the universe at ... one in 10 to the power of 100 billion" (p. 147).

Patterson (2007) relates that "Sir Fred Hoyle has offered the analogy of a tornado passing through a junkyard and assembling a Boeing 747, 'nonsense of a high order' in his words" (p. 147). Patterson comments that "natural selection cannot be the mechanism that caused life to form from matter as it can only work on a complete living organism" (p. 147).

Responding to the frequent claim of evolutionists that, despite the minutely small probabilities in favor of the chance occurrence of life by naturalistic means, Patterson (2007) explains:

> Bradley and Thaxton calculated the formation of a 100-amino-acid protein assembling by random chance to be 4.9×10^{-191}. It is generally accepted that any event with a probability beyond 1 x 10-50 is impossible, so we must conclude that evolution, requiring thousands of times this amount of complexity, is not likely to occur even if the entire universe were full of organic precursors. Naturalism and materialism can offer no realistic method for the origin of life on earth. (p. 144)

The probability that life arose by naturalistic processes

acting on earth in the remote past is so incredibly small that some scientists have proposed that "life originated on another planet" (Patterson, 2007, p. 144). Of course, that "solution" is no solution at all, since it then demands an explanation for the origin of life on that other planet.

Summary

Denton (1985) provides a very good summary of the overall lack of evidentiary support for the Darwinian approach:

> The overriding supremacy of the myth [Darwinism] has created a widespread illusion that the theory of evolution was all but proved one hundred years ago and that all subsequent biological research—paleontological, zoological and in the newer branches of genetics and molecular biology—has provided ever-increasing evidence for Darwinian ideas. Nothing could be further from the truth.... His general theory, that all life on earth had originated and evolved by a gradual successive accumulation of fortuitous mutations, is still, as it was in Darwin's time, a highly speculative hypothesis entirely without direct factual support and very far from that self-evident axiom some of its more aggressive advocates

would have us believe. (p. 77)

Criticisms of Darwinian theory are found on many fronts today. For example, Leith (1982) states:

> The main thrust of the criticism [of Darwinism] comes from within science itself.... These doubts are arising simultaneously from several independent branches of science. With a growth in the appreciation of the philosophy of science ... has come a doubt about whether Darwinism is, strictly speaking, scientific. Is the theory actually testable—as good theories must be? ... From within biology the doubts have come from scientists in half a dozen separate fields. Many paleontologists are unconvinced by the supposed gradualness of Darwinian evolution; they feel that the evidence points to abrupt change—or else to no change at all.... In the past ten years has emerged a new breed of biologists who are considered scientifically respectable, but who have their doubts about Darwinism. (pp. 10-11)

CHAPTER 4

CONCLUSIONS AND RECOMMENDATIONS

Conclusions

This dissertation has examined evidence from both the Scriptures and science with respect to the theory of evolution. Chapter 2 demonstrated overwhelmingly that the Bible teaches in both Old and New Testaments that God is Creator and that He created all things within a period of six days. No room is left for anyone to claim to believe the Bible and also to believe in naturalistic evolution. Theistic evolution is an unreasonable and untenable option.

Chapter 3 presented several lines of evidence that undermine the view that naturalistic evolution is a fact. Actually, there are many good reasons to reject the view that all things exist as a result of gradual, materialistic processes that occurred over billions of years. Numerous examples from science were introduced and examined to demonstrate the weaknesses of evolutionary theory.

Recommendations

Although an evolutionary approach to science will undoubtedly continue to be presented in schools, colleges, and universities because that view is currently dominant in academia, the following recommendations are offered to those in positions of decision-making authority.

First, when naturalistic evolution is taught, it should be presented as a **theory** and **not as a fact**.

Second, when a science course is introduced, that introductory material should include statements urging caution in making judgments and drawing conclusions. Examples of past errors in scientific theories should be provided in an effort at "full disclosure" of the limitations of science.

Third, when evolution is taught, the weaknesses of the theory should be introduced from a scientific perspective. That includes exposing students to the evidence that suggests alternatives to the prevailing viewpoint.

Fourth, while not necessarily introducing Biblical statements as evidence of creation, scientific evidence that favors "design"/"creation" should be included as an alternative to naturalism.

Under present circumstances in which our society is dominated by people who reject the Scriptural idea that God created all things by His great power, it is unlikely that the recommendations presented above will be implemented.

However, through the providential working of the Creator and through His amazing Word, sufficiently many people may eventually be turned to our gracious and merciful God to accomplish these recommendations to the glory of the Creator.

REFERENCES

Ackerknecht, E. H. 1973. Rudolf Virchow. *Encyclopedia Britannica*, 23: 35.

Allen, Katy Z., Linda Ruth Berg, Barbara Christopher, Jennie Dusheck, and Mark F. Taylor. 2007. *Life Science*. Orlando, FL: Holt, Rinehart and Winston.

Antevs, Ernst. 1957. Geological Tests of the Varve and Radiocarbon Chronologies. *Journal of Geology* 65.

Asimov, Isaac. 1977. *Beginning and end*. New York: Doubleday.

Behe, Michael. 1996. *Darwin's black box: The biochemical challenge to evolution*. New York: Touchstone.

Clark, A. H. 1930. *The new evolution: Zoogenesis*. Baltimore, MD: Williams and Wilkins.

Corner, E. J. H. 1961. Evolution. In A. M. MacLeod and L. S. Cobley (eds.), *Contemporary Botanical Thought*, pp. 95-114. Edinburgh: Oliver and Boyd

Darwin, Charles. 1859. *On the origin of species by means of natural selection*. London: John Murray.

Davies, Paul. 1988. *The cosmic blueprint*. New York: Simon & Schuster.

Dawkins, Richard. 1987. *The blind watchmaker*. New York: Norton.

Dembski, William. 2004. *The design revolution: Answering the toughest questions about intelligent design*. Downers Grove, IL: InterVarsity Press.

Dembski, William, and James Kushiner, eds. 2001. *Signs of intelligence*. Grand Rapids, MI: Baker.

Denton, Michael. 1985. *Evolution: A theory in crisis*. Bethesda, MD: Adler & Adler.

Dobzhansky, Theodorius, *Evolution, Genetics, and Man*. 1955. New York: John Wiley and Sons.

Feldkamp, Susan, ed. 2002. *Modern biology*. Austin, TX: Holt, Rinehart and Winston.

Geisler, Norman. 1999. *Baker encyclopedia of Christian apologetics*. Grand Rapids, MI: Baker.

Geisler, Norman L., and Frank Turek. 2004. *I don't have enough faith to be an atheist*. Wheaton, IL: Crossway Books.

Geisler, Norman, and Paul Hoffman, eds. 2001. *Why I am a Christian: Leading thinkers explain why they believe*. Grand Rapids, MI: Baker.

Geisler, Norman, and Peter Bocchino. 2001. *Unshakable foundations*. Minneapolis, MN: Bethany.

Gould, Stephen J. 1977. Evolution's Erratic Pace. *Natural History* 86 (May 1977): 13-14.

Gurin, Joel. 1981. *The Sciences* 21(4), April 1981: 16-19.

Hardin, Garrett. 2010. *Nature and man's fate*. Whitefish, MT: Kessinger Publishing.

Hawking, Stephen W. 1988. *A brief history of time*. New York: Bantam.

Heeren, Fred. 2000. *Show me God*. Wheeling, IL: Daystar.

Henry, Matthew. 1828. *An exposition of the Old and New Testament*. Volume 1. London: Joseph Ogle Robinson.

Heribert-Nilsson, Nils. 1953. *The synthetic origin of species*. Lund, Sweden: Verlag CWK Gleerup.

Hotton, Nicholas, III. 1968. *The evidence of evolution*. New York: American Heritage Publishing Co.

Jastrow, Robert. 1978. *God and the astronomers*. New York: Norton.

Johnson, Phillip E. 2000. *The wedge of truth*. Downers Grove, IL: InterVarsity Press.

Johnson, Phillip E. 1993. *Darwin on trial*. Downers Grove, IL: InterVarsity Press.

Keith, M. L., & G. M. Anderson. 1963. Radiocarbon Dating: Fictitious Results with Mollusk Shells. *Science* 141: 634-637.

Kelso, A. J. 1974. Origin and Evolution of the Primates. *Physical Anthropology*. New York: J. B Lippincott.

Kennedy, D. James. 1997. *Skeptics answered*. Sisters, OR: Multnomah.

Leith, Brian. 1982. *The descent of Darwinism: A handbook of doubts about Darwinism*. Toronto, Ontario, Canada: William Collins & Sons..

Lewontin, Richard. Billions and Billions of Demons. *The New York Review of Books* (January 9, 1997. Retrieved from http://www.nybooks.com/articles/archives/1997/jan/09/billions-and-billions-of-demons/ on October 3, 2015..

Lipson, H. S. 1980. A Physicist Looks at Evolution. *Physics Bulletin* 31: 138.

Mader, Sylvia S. 2010. *Concepts of Biology*. Second Edition. New York: McGraw-Hill.

Montagu, Ashley, ed. 1984. *Science and Creationism*. New York: Oxford University Press.

Morris, Henry M. 2002. *The Biblical basis for modern science*. Green Forest, AR: Master Books.

NABT Board of Directors. 2011. *NABT Position statement on teaching evolution*. Retrieved from http://nabt.org/websites/institution/index.php?p=92 on September 10, 2015.

National Academy of Sciences. 1998. *Teaching about evolution and the nature of science*. Washington, DC: National Academy Press.

Nelson, Byron. 1967. *After its kind*. Minneapolis, MN: Bethany Fellowship.

Otto, James H., Albert Towle, and James V. Bradley. 1981. *Modern Biology*. New York: Holt, Rinehart and Winston.

Patterson, Roger. 2007. *Evolution exposed: Biology*. Hebron, KY: Answers in Genesis.

Raup, David M. 1979. Conflicts Between Darwin and Paleontology. *Field Museum of Natural History Bulletin*, 50(1), January 1979: 22-29.

Reed, Charles A. 1959. Animal Domestication in the Prehistoric Near East. *Science* 130: 1629-1639.

Riegle, David D. 1971. *Creation or evolution?* Grand Rapids, MI: Zondervan Publishing House.

Ross, Hugh. 1995. *The Creator and the cosmos*. Colorado Springs, CO: NavPress.

Sagan, Carl. 1980. *Cosmos*. New York: Random House.

Sarfati, Jonathan. 1999. *Refuting evolution: A handbook for students, parents, and teachers countering the latest arguments for evolution*. Green Forest, AR: Master Books.

Sarfati, Jonathan. 2002. *Refuting evolution 2*. Green Forest, AR: Master Books.

Shute, Evan. 1961. *Flaws in the theory of evolution*. Grand Rapids, MI: Baker Book House.

Simpson, George Gaylord. 1953. *The major features of evolution*. New York: Columbia University Press.

Smallwood, William L., and Edna R. Green. 1971. *Biology*. Morristown, NJ: Silver Burdett.

Stahl, Barbara J. 1974. *Vertebrate history: Problems in evolution*. New York: McGraw- Hill.

Strobel, Lee. 2000. *The case for faith*. Grand Rapids, MI: Zondervan.

The Holy Bible, New King James Version. 1994. Nashville, TN: Thomas Nelson.

Thompson, Adell. 1983. *Biology, zoology, and genetics: Evolution model vs. creation model*. Washington, DC: University Press of America.

Thompson, Bert. 1986. *Is Genesis myth?* Montgomery, AL: Apologetics Press.

Vine, W. E., Merrill F. Unger, and William White, Jr. 1985. *Vine's Expository Dictionary of Biblical Words*. Nashville, TN: Thomas Nelson Publishers.

Ward, Peter, and Donald Brownlee. 2000. *Rare Earth*. New York: Copernicus.

Watson, D. M. S. 1929. Adaptation. *Nature*, 124:233.

Watson, Lyall. 1982. The Water People. *Science Digest*, 90 (May 1982): 44.

Wells, Jonathan. 2000. *Icons of evolution: Science or myth? Why much of what we teach about evolution is wrong*. Washington, DC: Regnery.

White, David L., and Thomas N. Thrasher. 1993. Videotape of a public debate on evolution held on December 2, 1993, in Decatur, AL.

Wysong, R. L. 1976. *The creation-evolution controversy*. Midland, MI: Inquiry Press.

Yockey, Hubert. 1992. *Information theory and molecular biology*. Cambridge, NY: Cambridge University Press.

Yockey, H. P. 1977. A calculation of the probability of spontaneous biogenesis by information theory. *Journal of Theoretical Biology*, 67:377-398.

THRASHER PUBLICATIONS
1705 Sandra Street S.W.
Decatur, AL 35601-5457
Email: thomas.thrasher@att.net

Bogard—McPherson Debate on miraculous healing
   Ben M. Bogard (Baptist) and Aimee Semple McPherson (Foursquare)
Calhoun—Kurfees Discussion on instrumental music in the worship
   H. L. Calhoun (Christian) and M. C. Kurfees (Christian)
Donahue-Thrasher Exchange on eternal life as a present possession
   Patrick T. Donahue (Christian) and Thomas N. Thrasher (Christian)
Falls—Franklin Debate on Holy Spirit Baptism & Gifts of the Spirit
   Drew E. Falls (Christian) and Ben J. Franklin (Charismatic)
Falls—Speakman Debate on Miracles
   Drew E. Falls (Christian) and Lummie Speakman (Pentecostal)
Falls—Storment Debate on the coverings of 1 Corinthians 11
   Drew E. Falls (Christian) and Keith Storment (Christian)
Falls—Welch Debate on the coverings of 1 Corinthians 11
   Drew E. Falls (Christian) and D. L. Welch (Pentecostal)
Garrett-Thrasher Debate on the Great Commission
   Eddie K. Garrett (Primitive Baptist) and Thomas N. Thrasher (Christian)
Madrigal—Mayo Debate on the necessity of water baptism
   Dan Mayo (Baptist) and John R. Madrigal (Christian)
McCay—Porter Debate on the communion cup
   G. Earl McCay (Christian) and Rue Porter (Christian)
Must We Keep the Sabbath Today?
   Carrol R. Sutton (Christian)
O'Neal—Hicks Debate on church-sponsored recreational activities
   Thomas G. O'Neal (Christian) and Olan Hicks (Christian)
Porter—Dugger Debate on the Sabbath and the Lord's Day
   W. Curtis Porter (Christian) and A. N. Dugger (Church of God–7th Day)
Rejecting Naturalistic Theories of Origins: Scientific and Scriptural Arguments. Thomas N. Thrasher (Christian)
Scambler—Langley Debate on the truth of Christianity
   T. H. Scambler (Christian) and J. S. Langley (Rationalist)
Sutton—Woods Debate on Congregational Benevolence
   Carrol Ray Sutton (Christian) and Guy N. Woods (Christian)

Tant—Frost Debate on instrumental music and societies
 J. D. Tant (Christian) and W. G. Frost (Christian)

Tant—Harding Debate on rebaptism
 J. D. Tant (Christian) and James A. Harding (Christian)

Tant—Smith Debate on Alexander Campbell's baptism
 J. D. Tant (Christian) and C. A. Smith (Baptist)

Thrasher—Barr Debate on the identity of the New Testament church
 Vernon L. Barr (Baptist) and Thomas N. Thrasher (Christian)

Thrasher—Coleman Debate on the Lord's Supper
 Pat S. Coleman (Pentecostal) and Thomas N. Thrasher (Christian)

Thrasher—Davis Debate: Will Everyone Be Eternally Saved?
 Myles Davis (Universalist) and Thomas N. Thrasher (Christian)

Thrasher—Forsythe Debate on the church of Christ
 Richard W. Forsythe (Pentecostal) and Thomas N. Thrasher (Christian)

Thrasher—Garrett Debate on unconditional salvation and apostasy
 Eddie K. Garrett (Primitive Baptist) and Thomas N. Thrasher (Christian)

Thrasher—Green Debate on the Christian and civil government
 Ken Green (Christian) and Thomas N. Thrasher (Christian)

Thrasher—Martignoni Debate: Was Peter the First Pope?
 John Martignoni (Roman Catholic) and Thomas N. Thrasher (Christian)

Thrasher—Maxey Debate on eternal punishment
 Al Maxey (Christian) and Thomas N. Thrasher (Christian)

Thrasher—Mayo Debate on the impossibility of apostasy
 Dan Mayo (Baptist) and Thomas N. Thrasher (Christian)

Thrasher—Miller Debate on Bible classes and women teachers
 E. H. Miller (Christian) and Thomas N. Thrasher (Christian)

Thrasher—Owens Debate on everlasting punishment for the wicked
 Lester Owens (Seventh Day Adventist) and Thomas N. Thrasher (Christian)

Thrasher—Waters Debate on divorce and remarriage
 Robert Waters (Christian) and Thomas N. Thrasher (Christian)

Thrasher—Welch Debate on the formula of words used in baptism
 D. L. Welch (Pentecostal) and Thomas N. Thrasher (Christian)

Thrasher—White Debate on Creation versus Evolution
 David L. White (Evolutionist) and Thomas N. Thrasher (Creationist)

Warnock—Williams Discussion on weddings and funerals in the meetinghouse
 Weldon E. Warnock (Christian) and Ralph D. Williams (Christian)

www.ingramcontent.com/pod-product-compliance
Lightning Source LLC
Chambersburg PA
CBHW070425180526
45158CB00017B/759